**Chaymae Laoufi**
**Ahmed Abbou**
**Mohammed Akherraz**

**Comparative study between different speed controller techniques**

AF153165

Chaymae Laoufi
Ahmed Abbou
Mohammed Akherraz

# Comparative study between different speed controller techniques

**Performance improvement of several commands of an induction machine**

**LAP LAMBERT Academic Publishing**

**Impressum / Imprint**

Bibliografische Information der Deutschen Nationalbibliothek: Die Deutsche Nationalbibliothek verzeichnet diese Publikation in der Deutschen Nationalbibliografie; detaillierte bibliografische Daten sind im Internet über http://dnb.d-nb.de abrufbar.

Alle in diesem Buch genannten Marken und Produktnamen unterliegen warenzeichen-, marken- oder patentrechtlichem Schutz bzw. sind Warenzeichen oder eingetragene Warenzeichen der jeweiligen Inhaber. Die Wiedergabe von Marken, Produktnamen, Gebrauchsnamen, Handelsnamen, Warenbezeichnungen u.s.w. in diesem Werk berechtigt auch ohne besondere Kennzeichnung nicht zu der Annahme, dass solche Namen im Sinne der Warenzeichen- und Markenschutzgesetzgebung als frei zu betrachten wären und daher von jedermann benutzt werden dürften.

Bibliographic information published by the Deutsche Nationalbibliothek: The Deutsche Nationalbibliothek lists this publication in the Deutsche Nationalbibliografie; detailed bibliographic data are available in the Internet at http://dnb.d-nb.de.

Any brand names and product names mentioned in this book are subject to trademark, brand or patent protection and are trademarks or registered trademarks of their respective holders. The use of brand names, product names, common names, trade names, product descriptions etc. even without a particular marking in this work is in no way to be construed to mean that such names may be regarded as unrestricted in respect of trademark and brand protection legislation and could thus be used by anyone.

Coverbild / Cover image: www.ingimage.com

Verlag / Publisher:
LAP LAMBERT Academic Publishing
ist ein Imprint der / is a trademark of
OmniScriptum GmbH & Co. KG
Heinrich-Böcking-Str. 6-8, 66121 Saarbrücken, Deutschland / Germany
Email: info@lap-publishing.com

Herstellung: siehe letzte Seite /
Printed at: see last page
**ISBN: 978-3-659-78183-4**

Copyright © 2015 OmniScriptum GmbH & Co. KG
Alle Rechte vorbehalten. / All rights reserved. Saarbrücken 2015

# TABLES OF CONTENTS

# NOMENCLATURE

$i_{s\alpha}, i_{s\beta}$ : Stator currents; $\alpha$- and $\beta$- axis components

$V_{s\alpha}, V_{s\beta}$ : Stator voltages; $\alpha$- and $\beta$- axis components

$\Phi_{rd}, \Phi_{rq}$ : Rotor flux; d- and q- axis components

$\Phi_r$ : Rotor flux magnitude

$L_s, L_r, M$ : Stator, rotor and mutual inductances

$R_s, R_r$ : Stator, rotor resistance

$C_{em}$ : Electromagnetic torque

$T_s, T_r$ : Stator, rotor time constant

$\theta_s$ : Synchronous reference frame position

$\omega_s$ : Angular speed

$\omega_r$ : Mechanical rotor speed

$\sigma$ : Total leakage factor

p : Number of pole pairs

J : Total inertia moment

f : Total viscous friction coefficient

# LIST OF FIGURES

# LIST OF TABLES

# INTRODUCTION

The induction machine is widely used in industrial applications considering its performance. It is increasingly used for performing controls by replacing the DC motor. The main constraint of the control of this machine is related to the absence of the decoupling between the flux and the torque.

The indirect field-oriented control and direct torque control drives are the most powerful commands used in industry.

In this work, two studies have been performed to improve the performance of these commands.

The first study consists of a comparison between different speed controller techniques to improve the performance of indirect field-oriented control and the second study presents a solution for elimination of stator resistance variation effect in direct torque control to improve the performance of this command.

In fact, the dynamic characteristics of indirect field-oriented control drive are complex, nonlinear and coupled. Also, it is sensitive to the disturbance of the induction machine parameters.

In the order to limit these drawbacks, several intelligent controllers such as sliding mode controller (SMC), self tuning fuzzy logic controller (PIST) and fuzzy logic controller (FLC) have received much attention in recent years to controlling this drive.

The sliding mode is an effective control strategy for nonlinear systems with uncertainties [1]. Its principle is based on the definition of a surface called sliding surface depending on system states so that it is attractive. The synthesized global control consists of two terms: the first allows the stat trajectory to approach this surface and the second maintaining and sliding along it towards the origin of the phase plane [2]. It is characterized by good robustness, fast response time and disturbance rejection. However, one of the drawbacks of this controller is the chattering phenomenon caused by the discontinue control action.

The fuzzy logic controller presents the high performances of speed tracking. However, this controller is insufficient to deal with systems subjected to server perturbation because its gains are fixed [3]-[6]. So, to improve the limited performances of the fuzzy logic controller, in case of the disturbance parameters, the self tuning fuzzy logic controller have been developed [7]-[10].

In this work, the design of the sliding mode and the self tuning fuzzy logic controllers are proposed. The comparison of the different controllers is established in case of two methods. In first method, the different controllers are applied to adapt the error between the actual rotor speed and the reference speed. The second method consists to force the systems to follow a reference model by comparing the rotor speed with the output of this reference model [10]-[11].

Otherwise, In spite of its simplicity, direct torque control (DTC) provides a fast and precise torque response. However, one of the drawbacks of this drive is the sensitivity to the variation of the stator resistance.

In fact, the stator resistance is used to estimate the stator flux. So, any variation of stator resistance due to changes in temperature and frequency degrades the performances of this control strategy and affects its stability at low speeds.

In order to limit this drawback, several methods have been developed such as estimation of the stator resistance and its compensation which presents some limits, among which the necessity to determine the analytic model to calculate the stator current reference and the dependence on stator inductances [13], [14].

In this work, the self tuning fuzzy logic controller has been developed to compensate the effect of stator resistance variation. The principle of this controller consists to

7

control online the gains of the classical PI controller by using a fuzzy logic adapter and adjustment of these gains when a variation of stator resistance is detected [10], [11].

The simulation results show the performances and limits of the proposed controllers. The self tuning fuzzy logic controller is advantageous in term of response time while the sliding mode controller and fuzzy logic controller present a high capacity to reject the disturbance. Also, high effectiveness of the self tuning fuzzy logic controller to reject the disturbance of stator resistance and maintain the stability of the direct torque control has been shown.

# I.    Modeling of induction machine

The induction machine (asynchronous machine) is the most used in all industrial applications considering its ease of implementation, its good efficiency and its excellent reliability. But one of its drawbacks is the physical complexity related to the electromagnetic interaction between the stator and the rotor. This type of machine has three main parts, rotor, stator, and enclosure. The stator and rotor do the work, and the enclosure protects the stator and rotor. The modeling of the induction machine is based on a number of assumptions:

- perfect symmetry ;
- Assimilation to a rotating machine with three-phases in the stator and three-phases in the rotor;
- Sinusoidal distribution of the magnetic field along the air gap;
- Negligible saturation and losses in the magnetic circuit;
- Negligence of the influence of the skin effect and overheating of conductors.

The Figure 1 gives a representation of three-phase induction machine where:

- (a,b,c)   : represent the indices of the three phases on the stator;
- (A,B,C) : represent the indices of the three phases on the rotor.

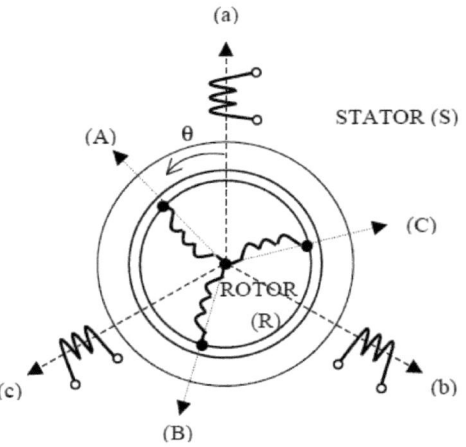

(a)

(A)  θ

STATOR (S)

(C)

ROTOR
(R)

(c)

(b)

(B)

Figure. 1. Modeling of the induction machine

In the case of the reference $(\alpha,\beta)$, equations of the machine are given in general as shown below [12]:

$$\frac{di_{s\alpha}}{dt} = -\left(\frac{1}{\sigma T_s} + \frac{(1-\sigma)}{\sigma T_r}\right) i_{s\alpha} + w_s i_{s\beta} \frac{(1-\sigma)}{\sigma M T_r} \phi_{r\alpha} + \frac{(1-\sigma)}{\sigma M} w\phi_{r\beta} + \frac{1}{\sigma L_s} V_{s\alpha} \qquad (1)$$

$$\frac{di_{s\beta}}{dt} = -w_s i_{s\alpha} + \left(\frac{1}{\sigma T_s} + \frac{(1-\sigma)}{\sigma T_r}\right) i_{s\beta} - \frac{(1-\sigma)}{\sigma M} w\phi_{r\alpha} + \frac{(1-\sigma)}{\sigma M T_r} w\, \phi_{r\beta} + \frac{1}{\sigma L_s} V_{s\beta} \qquad (2)$$

$$\frac{d\phi_{r\alpha}}{dt} = \frac{M}{T_r} i_{s\alpha} - \frac{1}{T_r} \phi_{r\alpha} + (w_s - w)\phi_{r\beta} \qquad (3)$$

$$\frac{d\phi_{r\beta}}{dt} = \frac{M}{T_r} i_{s\beta} - (w_s - w)\phi_{r\alpha} - \frac{1}{T_r} \phi_{r\beta} \qquad (4)$$

$$C_{em} = \frac{pM}{L_r}\left(\phi_{r\alpha} i_{s\beta} - \phi_{r\beta} i_{s\alpha}\right) \qquad (5)$$

# II. Control techniques of the induction machine

## II.1 Indirect rotor field-oriented control

Indirect rotor field-oriented control was developed for the purpose of decoupling the torque and the flux. This decoupling allows for a very fast response of torque. The principle of field-oriented control is to represent the dynamic model of the induction machine in the rotating reference with the rotor flux (Figure 2).

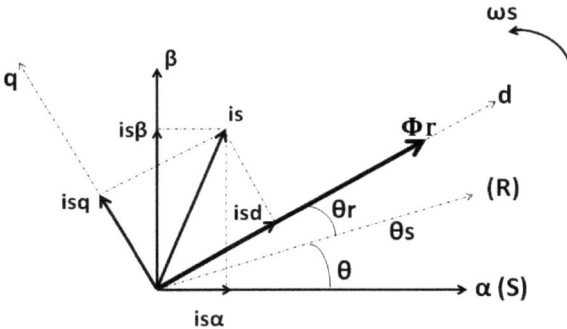

Figure. 2. Illustration of the rotor flux orientation

The alignment of the rotor flux $\Phi_r$ on the d-axis of the rotating reference implies the following:

$$\Phi_{rd} = \Phi_r \text{ and } \Phi_{rq} = 0 \qquad (6)$$

The principle of this command consists to enforce the current of the induction machine to follow the reference currents $i_{sd\_ref}$ and $i_{sq\_ref}$ which are perfectly decoupled.

In the present work, the indirect rotor field-oriented command controlled by current (IRFOCC) is used. This technique is characterized by the absence of the rotor flux loop. So this flux is not measured and not estimated. Therefore, sensors and observers are not required [12].

## II.1.2 Structure of the indirect rotor field-oriented control

In the case of this command, the rotor flux $\Phi_r$ is aligned with the d-axis of the rotating reference. This implies:

$$\Phi_{rd} = \Phi_r \text{ and } \Phi_{rq} = 0 \tag{7}$$

The equations of the induction machine with rotor flux oriented are given by the following:

If $\Phi_r$ is constant:

$$\Phi_r = Mi_{sd} \tag{8}$$

$$C_{em} = p \frac{M}{L_r} \Phi_r i_{sq} \tag{9}$$

The magnitude of the rotor flux $\Phi_r$ is determined only by the direct component of the stator current $i_{sd}$. And the electromagnetic torque $C_{em}$ is determined by the quadrature component of the stator current $i_{sq}$.

Since the flux is set at its reference value and maintained constant, the electromagnetic torque expression is given as follow:

$$C_{em} = K\phi_r i_{sq} \tag{10}$$

This equation is similar to the one of DC motor, where the torque depends only on the quadrature component of the stator current $i_{sq}$, if the flux $\Phi_r$ is kept constant.

So, finally we can see that the problem of coupling is removed between the two axes direct (d) and quadrature (q).

The structure of the speed drive (IRFOCC) comprises a conventional regulator of speed PI is shown by Figure 3:

The induction machine is controlled by current through an inverter whose the switching logic is provided by three hysteresis controllers.

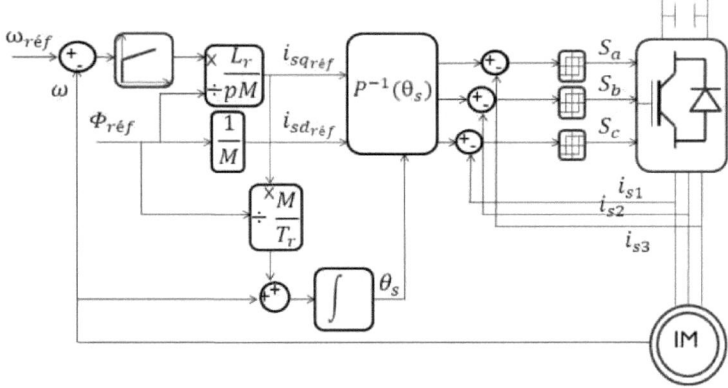

Figure. 3. Structure of the speed drive (IRFOCC)

In fact, the gains of the PI controller depend on machine parameters (rotor resistance $R_r$, inertia J) and external parameters (load torque $C_r$). Therefore any disturbance of these parameters directly influences these gains. The gains of the PI controller are given by the following equations:

$$K_\Omega = \frac{R_r}{4\sigma L_r} \qquad (11)$$

$$\tau_\Omega = \frac{J}{f} \qquad (12)$$

In fact, the machine parameters are variable and depend on the experimental conditions.

## II.2 Direct torque control

The structure of the direct torque control, as shown in Figure 4, compounds hysteresis torque and flux controllers and PI speed controller.

The expression of the flux in the reference $(\alpha,\beta)$ is given by the following equation:

$$\Phi_{s\alpha} = \int V_{s\alpha} - R_s i_{s\alpha} + \Phi_{s\alpha 0} \tag{13}$$

$$\Phi_{s\beta} = \int V_{s\beta} - R_s i_{s\beta} + \Phi_{s\beta 0} \tag{14}$$

The magnitude and the position of the flux are obtained by:

$$\Phi_s = \sqrt{\Phi_{s\alpha}^2 + \Phi_{s\beta}^2} \tag{15}$$

$$\theta_s = \tan^{-1}(\frac{\Phi_{s\beta}}{\Phi_{s\alpha}}) \tag{16}$$

The electromagnetic torque is given by:

$$C_{em} = \frac{3}{2} p (\Phi_{s\alpha} i_{s\beta} - \Phi_{s\beta} i_{s\alpha}) \tag{17}$$

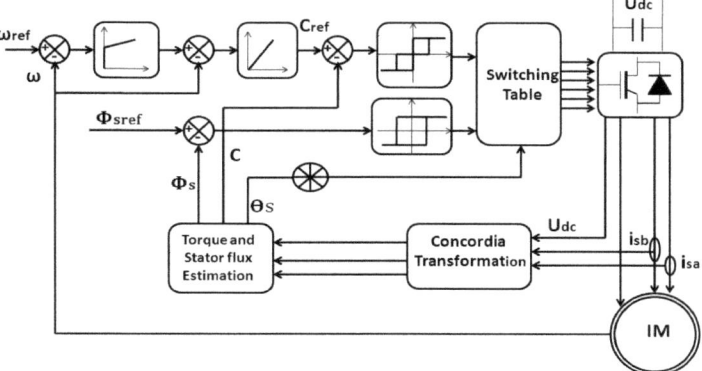

Figure 4. Structure of DTC

Direct torque control is based on the directly determination of the sequence of control applied to the switches of a tension inverter. This choice is generally based on the use of hysteresis regulators, whose function is to control the state of the system, and to modify the amplitude of the stator flux and the electromagnetic torque [14].

From the equations (13)-(14) and as the voltage drop due to the stator resistance can be neglected for high speeds. The expression of the stator flux becomes:

$$\overrightarrow{\Phi s}(t) \simeq \int_0^t \overrightarrow{V_s} + \overrightarrow{\Phi_{s0}} \tag{18}$$

During one period of sampling Te, the tension vector applied to the induction machine remains constants. Then this equation can be written as follows:

$$\overrightarrow{\Phi s}(k+1) \simeq \overrightarrow{\Phi s}(k) + \overrightarrow{V_s} T_e \tag{19}$$

15

Or even:

$$\Delta\overrightarrow{\Phi_s} \simeq \overrightarrow{V_s}T_e \tag{20}$$

This equations implies that the extremity of the vector $\overrightarrow{\Phi s}$ moves in a straight line whose direction is given by the implied tension vector $\overrightarrow{V_s}$, as shown in the Figure 5.

So, with a choice of a correct sequence of tension vector $\overrightarrow{V_s}$ in successive intervals of time of length Te, we can force the extremity of the flux vector $\overrightarrow{\Phi_s}$ to follow the desired trajectory. Therefore, to increase the stator flux, it suffices to apply a tension vector that is collinear in its direction and vice versa.

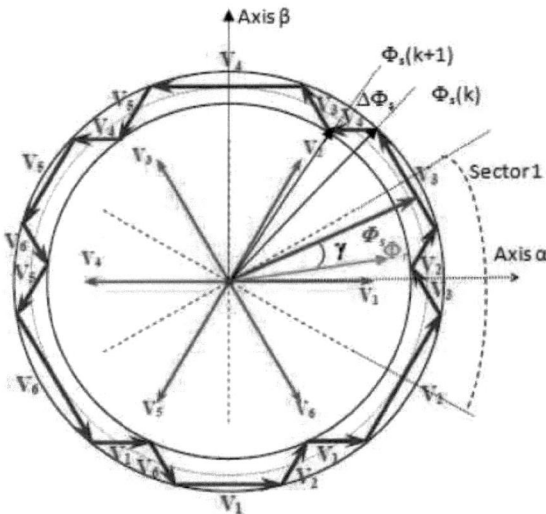

Figure.5. Definition of stator flux increment and spatial positions of the voltage vectors keeping the flux inside the strip of hysteresis [12], [14].

The projections of the flux error $\overrightarrow{\Delta\Phi_s}$ on the direction of stator flux and on a perpendicular direction, as shown in Figure 6, allow to act respectively on the magnitude of the stator flux (component $\Delta\Phi_{sf}$) and on the electromagnetic torque (component $\Delta\Phi_{sc}$).

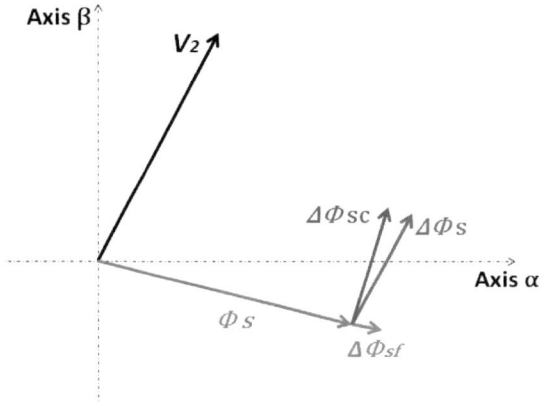

Figure. 6. Components of the flux error in case of the application of the tension vector $\overrightarrow{V_2}$

The electromagnetic torque, as shown by the following equation, is proportional to the vector product between the vectors of stator flux and rotor flux.

$$C_{em} = k\left(\overrightarrow{\Phi_s} \wedge \overrightarrow{\Phi_r}\right) = \frac{3}{2} p \frac{M}{\sigma L_s L_r} \Phi_s \Phi_r \sin \gamma \qquad (21)$$

So, the torque depends on the magnitude of the two vectors, stator flux $\overrightarrow{\Phi_s}$ and rotor flux $\overrightarrow{\Phi_r}$, and their relative position $\gamma$. If we succeed in perfectly controlling of the stator flux $\overrightarrow{\Phi_s}$ (from the tension vector $\overrightarrow{V_s}$) in module and in position, we can thereafter control the magnitude and the position of the rotor flux $\overrightarrow{\Phi_r}$ and therefore the torque [12], [14].

The DTC strategy uses the stator resistance to estimate the flux and thereafter the torque. So, the variation of this resistance, due to heating of the machine during operation, considerably affects the control.

In this work, a compensation solution of the stator resistance variation will be presented.

## III. Sliding Mode Controller

The principle of this controller consists to control the system by forcing speed error $(e_\omega)$ and its derivative $(\frac{de_\omega}{dt})$ to move towards a sliding surface. The sliding surface is a scalar function defined by the following equation:

$$S(e_\omega, \frac{de_\omega}{dt}, t) = 0 \tag{22}$$

Where the sliding variable is:

$$S(t) = \frac{de_\omega(t)}{dt} + \lambda e_\omega(t) \tag{23}$$

With $\lambda$ is positive constant.

The object of the control is to maintain the surface to zero. This last equation is a linear differential equation whose unique solution is $e_\omega(t) = 0$.

In case of the speed controller design, the surface speed regulation is given by:

$$S(\omega) = \omega - \omega_{ref} \tag{24}$$

$$\frac{dS(\omega)}{dt} = \frac{d\omega}{dt} - \frac{d\omega_{ref}}{dt} \tag{25}$$

By replacing the speed expression, we obtain:

$$\frac{dS(\omega)}{dt} = \frac{d\omega}{dt} - \frac{d}{dt}\left(\frac{Mp}{JL_r}\Phi_r i_{sq} - \frac{f}{J}\omega_{ref} - \frac{1}{J}C_r\right) \tag{26}$$

Now, we replace the current $i_{sq}$ by the control current $i_{sq\text{-ref}}$. Such the structure of the sliding mode controller is constituted by two parts, one concerning the exact linearization ($i_{sqeq}$) and the other is stabilizing ($i_{sqn}$).So by replacing the current $i_{sq}$= $i_{sqeq}$+ $i_{sqn}$, the equation (17) becomes:

$$\frac{dS(\omega)}{dt} = \frac{d\omega}{dt} - \frac{d}{dt}\left(\frac{Mp}{JL_r}\Phi_r i_{sqeq} + \frac{Mp}{JL_r}\Phi_r i_{sqn} - \frac{f}{J}\omega_{ref} - \frac{1}{J}C_r\right) \tag{27}$$

During the sliding mode and steady state, we have $S(\omega) = 0$ and thereafter :

$$\begin{cases} \dfrac{dS(\omega)}{dt} = 0 \\ i_{sqn} = 0 \end{cases} \tag{28}$$

From where, we derive the formula of the equivalent command $i_{sqeq}$ :

$$i_{sqeq} = \frac{JL_r}{p\,M\Phi_r}\left(\frac{f}{J}\omega_{ref} + \frac{1}{J}C_r\right) \tag{29}$$

19

During the convergence mode, the following condition must be verified:

$$S(\omega).\frac{dS(\omega)}{dt} = 0 \tag{30}$$

By replacing the expression of the $i_{sqeq}$ in the equation (18), we obtain:

$$S(\omega) = -\frac{M\,p}{J\,L_r}\Phi_{r-ref}i_{sqn} \tag{31}$$

By choosing the form of the discontinuous command, so we pose:

$$i_{sqn} = K_\omega \text{Sign}(S(\omega)) \tag{32}$$

So, the speed sliding mode controller is determined. The block diagram of this controller is showed in the Figure 7:

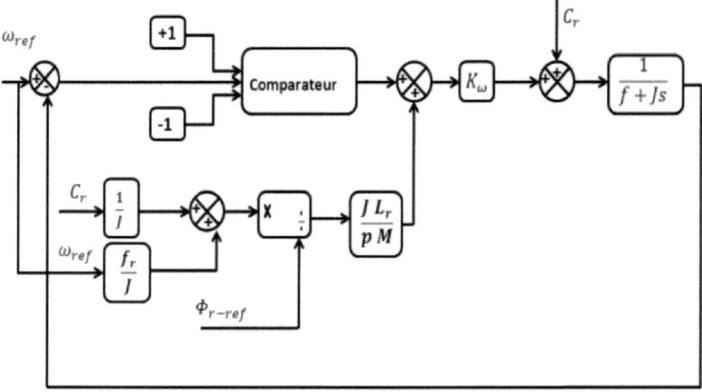

Figure. 7. Sliding mode controller for induction machine

## IV. Self Tuning Fuzzy Logic Controller

The block diagrams of the proposed methods are shown in the Figure 8 and Figure 10. The principle of the first method is to control the gains of a conventional PI in real time with the help of fuzzy logic and adjustment of these gains when a change is detected. In fact, the rotor speed is compared with the reference speed to generate the error $e_\omega$, this error and its derivative $\frac{de_\omega}{dt}$ are injected in the adaptation mechanism compound by a fuzzy logic adapter to generate the adaptation factors $\Delta K_p$ and $\Delta K_i$. These are also injected in the PI controller to correct the gains Kp and Ki.

The new parameters of the PI controller are obtained by:

$$Kp_f = Kp_i + \Delta Kp \qquad (33)$$

$$Ki_f = Ki_i + \Delta Ki \qquad (34)$$

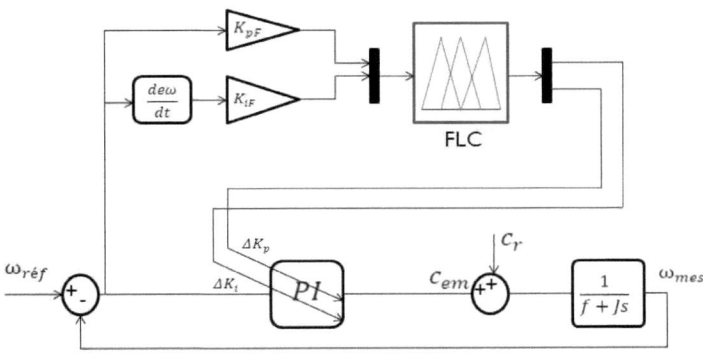

Figure. 8. Self-tuning fuzzy logic controller for induction machine

The principle of the second method is to force the system to follow a reference model. When a perturbation affects the parameters of the machine (rotor resistance, inertia) and external parameters (load torque) the adaptation mechanism corrects the gains of the PI controller to prevent the system to deviate from this reference model. In fact, the system composed by: the command IRFOCC, the inverter and the induction machine is equivalent to a low pass filter of the first order characterized by the following transfer function (Figure 9):

$$H_{BF} = \frac{1}{1+t_{rBF}\,s} \tag{35}$$

Where:

$s$    : The operator of Laplace.

$t_{rBF}$ : The closed loop response time determined by the response time of the response of the open-loop speed.

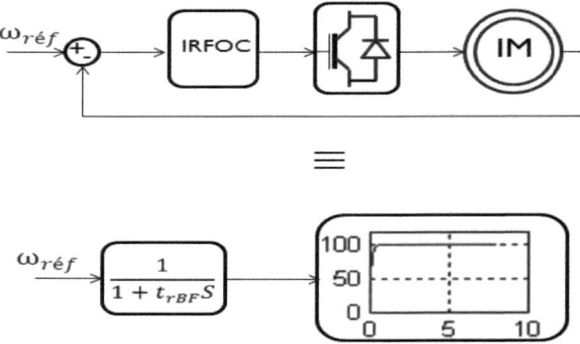

Figure. 9. Equivalence between the closed loop system and the low pass filter

The Figure 10 illustrates the principle of adaptation gains of the PI controller using the reference model. It is similar to the first method with the difference in the input of the adaptation mechanism. So the rotor speed is compared with the reference speed corresponding to the output of the reference model to generate the error $e_\omega$.

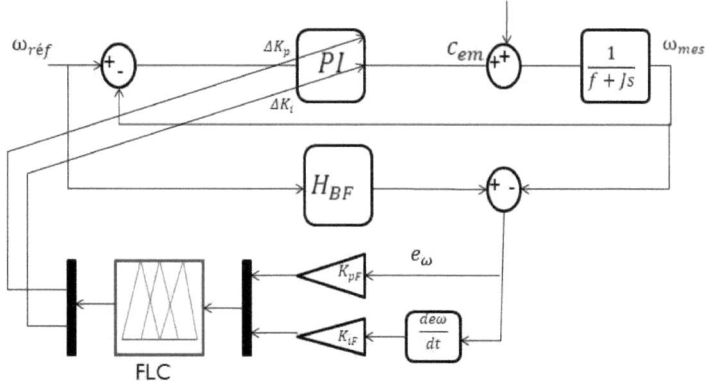

Figure. 10.  Adaptation gains of the PI controller using the reference model

### IV.1 Description of the Fuzzy Logic Controller

The inputs of the fuzzy logic controller are the speed error ($e_\omega$) and its derivative ($\frac{de_\omega}{dt}$). The fuzzy logic controller observes the error and updates the outputs ($K_p$, $K_i$).

The fuzzy subsets of input variables are defined as follows:

- GN : Big negative
- MN : Average negative
- PN : Small negative
- Z    : Zero
- PP  : Small positive
- MP : average positive
- GP : Big positive

The fuzzy subsets of the output variables are defined as follows:

- G : Big

- P : Small

The membership functions for inputs and outputs are defined in the interval [-1 1] as follows [8]:

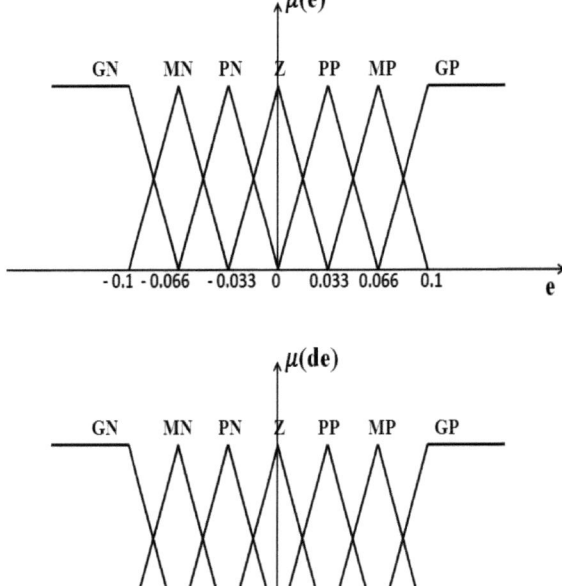

Figure. 11. Membership function for input variable

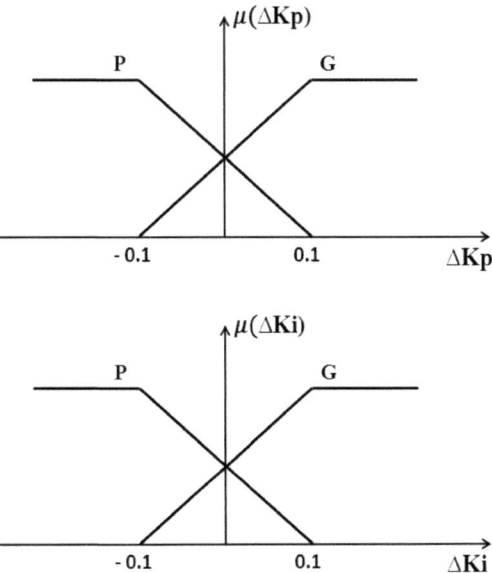

Figure. 12. Membership function for output variable

The bases rules used to calculate the output variables ΔKp and ΔKi are shown in the following tables 1 and 2.

Indeed, the fuzzy logic controllers used in this work is Mamdani type and the method of the defuzzification used is centroid method.

Table 1

Matrix inference to calculate ΔKp

|      | GN | MN | PN | Z | PP | MP | GP |
|------|----|----|----|---|----|----|----|
| GN   | G  | G  | G  | G | G  | G  | G  |
| MN   | P  | G  | G  | G | G  | G  | G  |
| PN   | P  | P  | G  | G | G  | P  | P  |
| Z    | P  | P  | P  | G | P  | P  | P  |
| PP   | P  | P  | G  | G | G  | P  | P  |
| MP   | P  | G  | G  | G | G  | G  | P  |
| GP   | G  | G  | G  | G | G  | G  | P  |

Table 2

Matrix inference to calculate ΔKI

|  | GN | MN | PN | Z | PP | MP | GP |
|---|---|---|---|---|---|---|---|
| GN | G | G | G | G | G | G | G |
| MN | G | G | P | P | P | G | G |
| PN | G | G | G | P | G | G | G |
| Z | G | G | G | P | G | G | G |
| PP | G | G | G | P | G | G | G |
| MP | G | G | P | P | P | G | G |
| GP | G | G | G | G | G | G | P |

## V.    Results of Simulation and Interpretation

## V.1 Simulation results of the first study (Indirect rotor field-oriented control)

In order to define the performances and limits of the proposed controllers: the sliding mode controller (SMC), the self tuning fuzzy logic controller (PIST), the fuzzy logic controller (FLC) and the classical PI controller, some perturbations on some parameters of the induction machine (rotor resistance, inertia) and the external parameters (load torque, reference speed) have been generated. Two tests have been performed, the first test relates to the low speeds and the second test involves high speeds. The value of listed parameters has been increased by 100% at t = 2 s and the value of the reference speed has been changed respectively from 20 rad/s to 50 rad/s and from 50 rad/s to 100 rad/s at t = 3 s. We applied these test for the two methods.

### V.1.1 First method simulations

The first test aims to evaluate the performances of the different controllers: sliding mode controller (SMC), self-tuning fuzzy logic controller (PIST), fuzzy logic controller and classical PI controller when the disturbance affects the rotor resistance.

As shown in Figure 13 and Figure 14, the sliding mode controller and fuzzy logic controller reject perfectly the perturbations compared to a self tuning fuzzy logic controller which in turn presents a better capacity to reject the disturbances than a conventional PI controller. In fact, the effect of the perturbations observed for a conventional PI is reduced by more than the half for a self tuning fuzzy logic controller. In terms of the response time, the self tuning fuzzy logic controller is characterized by a very small response time at startup and during the change of the speed compared to the other controllers. Also, we note that the proposed controllers, especially sliding mode controller and self tuning fuzzy logic controller, present the best performances even at low speeds. More speed quantitative performances are summarized in tables 3 and 4.

Case of the low speeds:

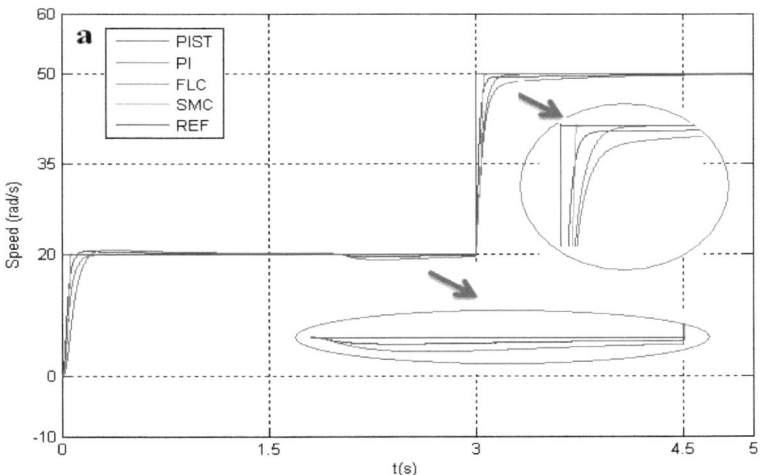

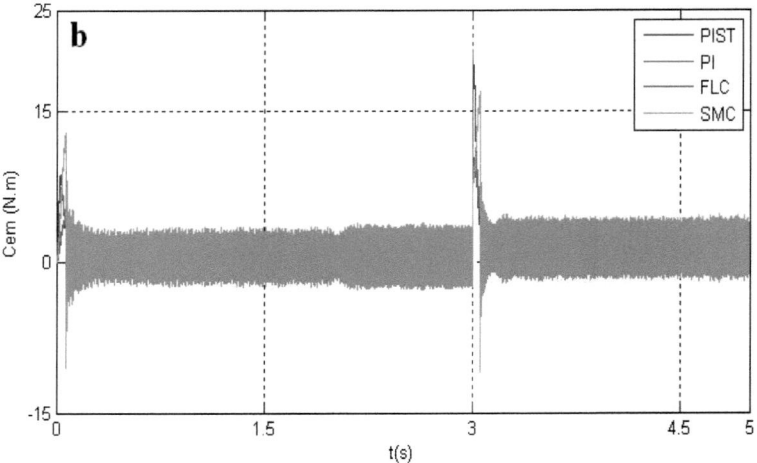

Figure 13. High tracking responses of the speed (a), the electromagnetic torque (b)
to change in rotor resistance (case of low speeds, 1st method).

## Case of the high speeds:

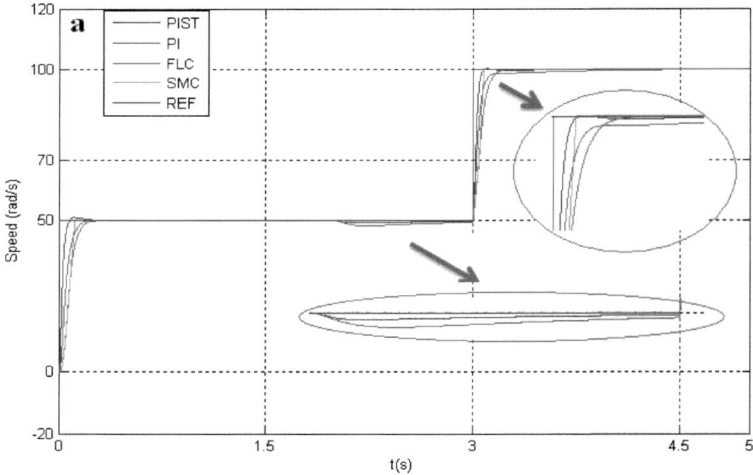

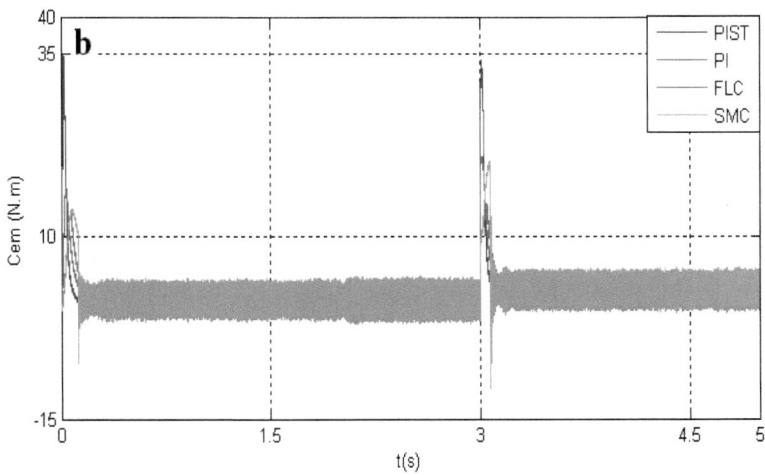

Figure. 14. High tracking responses of the speed (a), the electromagnetic torque (b) to change in rotor resistance (case of high speeds, 1$^{st}$ method).

Table 3

Quantitative performances of speed tracking in case of disturbance of the rotor resistance (case of low speeds, 1$^{st}$ method)

|  | Controllers | | | |
|---|---|---|---|---|
|  | PI | FLC | PIST | SMC |
| Response times (s) to attain 20 (rad/s) | 0.208 | 0.188 | 0.115 | 0.160 |
| Over shoot (%) In case of 20 (rad/s) | 0 | 0 | 1 | 0 |
| Response times (s) to attain 50 (rad/s) | 0.057 | 0.081 | 0.033 | 0.081 |
| Over shoot (%) In case of 50 (rad/s) | 0 | 0 | 0 | 0 |

Table 4

Quantitative performances of speed tracking in case of disturbance of

the rotor resistance (case of high speeds, 1$^{st}$ method)

| | Controllers | | | |
|---|---|---|---|---|
| | PI | FLC | PIST | SMC |
| Response times (s) to attain 50 (rad/s) | 0.174 | 0.252 | 0.084 | 0.186 |
| Over shoot (%) In case of 50 (rad/s) | 0 | 0 | 1.82 | 0 |
| Response times (s) to attain 100 (rad/s) | 0.045 | 0.087 | 0.033 | 0.066 |
| Over shoot (%) In case of 100 (rad/s) | 0 | 0 | 0.3 | 0 |

The second test consists to appraise the robustness of the proposed controllers towards the variations of the load torque. As shown in Figure 15 and Figure 16, a variation in the load torque and an acceleration of the speed are respectively applied at t = 2 s and t = 3 s. The results simulations show that the dynamic tracking of speed and the electromagnetic torque are better when using SMC and PIST controller and the conventional PI rejects less rapidly the perturbation. More speed quantitative performances are summarized in tables 5 and 6.

Case of the low speeds:

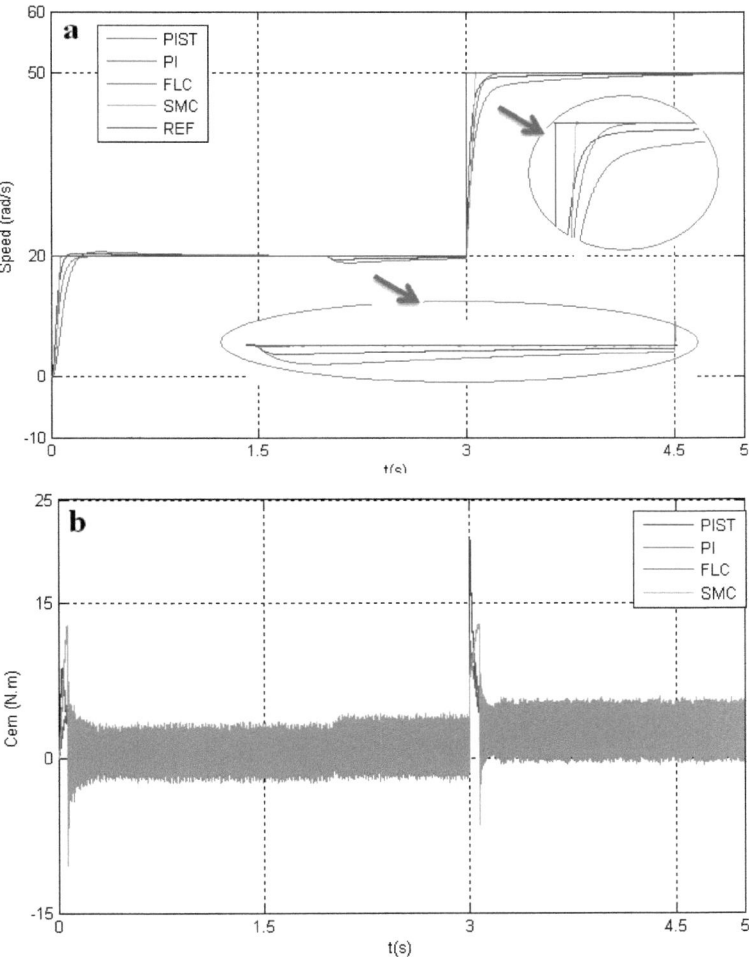

Figure. 15. High tracking responses of the speed (a), the electromagnetic torque (b) to change in load torque (case of low speeds, 1$^{st}$ method).

Case of the high speeds:

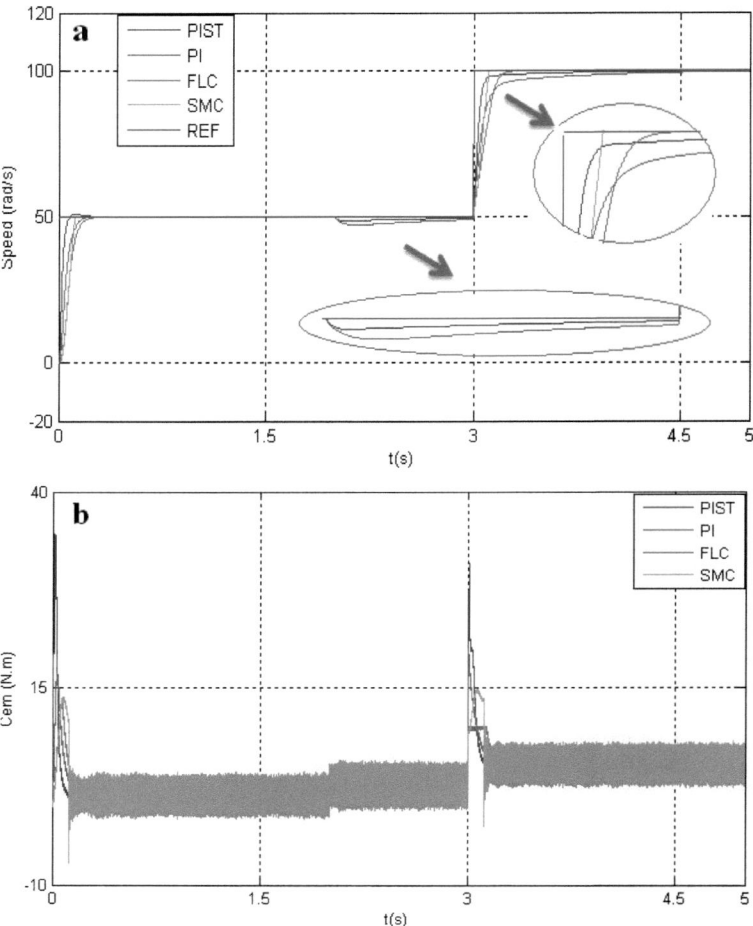

Figure. 16. High tracking responses of the speed (a), the electromagnetic torque (b)
to change in load torque (case of high speeds, 1st method).

Table 5

Quantitative performances of speed tracking in case of disturbance of

the load torque (case of low speeds, 1$^{st}$ method)

| | Controllers | | | |
|---|---|---|---|---|
| | PI | FLC | PIST | SMC |
| Response times (s) to attain 20 (rad/s) | 0.207 | 0.186 | 0.114 | 0.159 |
| Over shoot (%) In case of 20 (rad/s) | 0 | 0 | 1 | 0 |
| Response times (s) to attain 50 (rad/s) | 0.066 | 0.084 | 0.039 | 0.09 |
| Over shoot (%) In case of 50 (rad/s) | 0 | 0 | 0 | 0 |

Table 6

Quantitative performances of speed tracking in case of disturbance of

the load torque (case of high speeds, 1$^{st}$ method)

| | Controllers | | | |
|---|---|---|---|---|
| | PI | FLC | PIST | SMC |
| Response times (s) to attain 50 (rad/s) | 0.174 | 0.252 | 0.084 | 0.186 |
| Over shoot (%) In case of 50 (rad/s) | 0 | 0 | 0 | 0 |
| Response times (s) to attain 100 (rad/s) | 0.057 | 0.105 | 0.039 | 0.084 |
| Over shoot (%) In case of 100 (rad/s) | 0 | 0 | 0 | 0 |

33

The aim of a third test is to compare the performances of the three controllers; SMC, PIST and FLC; in the case where a disturbance affects the inertia of the machine. As the previous cases, the PIST has the best performance in terms of response time with a small overshoot when changing the speed at second t=3 s. Also, it can be seen that the change of the operating point affects less a dynamic tracking of speed and the electromagnetic torque for a sliding mode controller compared to the other controllers.

Case of the low speeds:

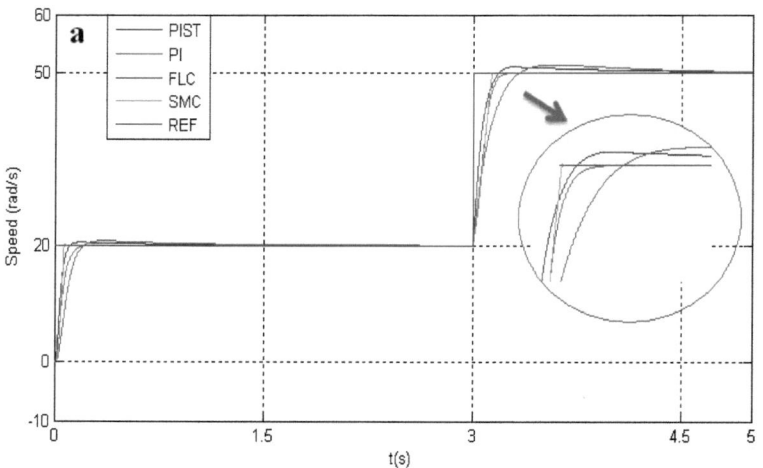

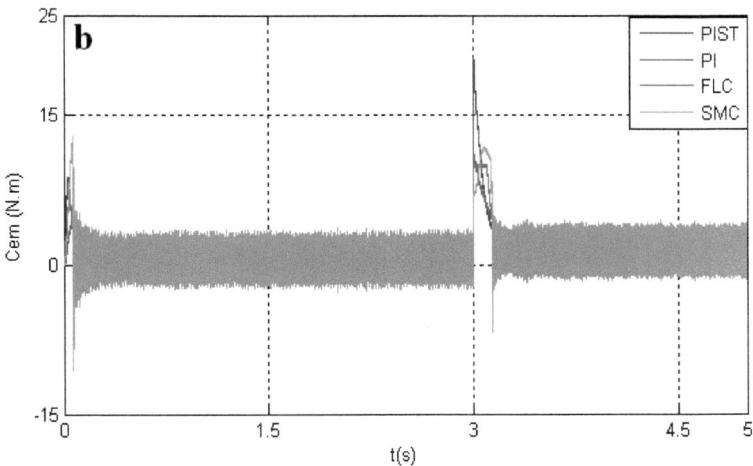

Figure. 17. High tracking responses of the speed (a) and of the electromagnetic torque (b)
to change in inertia (case of low speeds, 1$^{st}$ method).

Case of the high speeds:

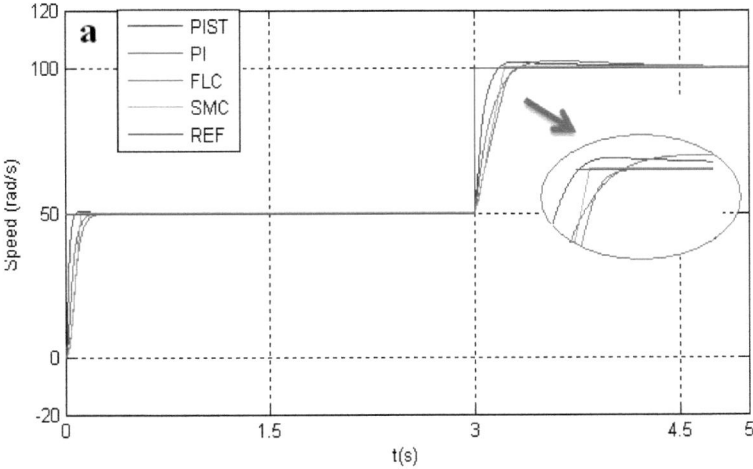

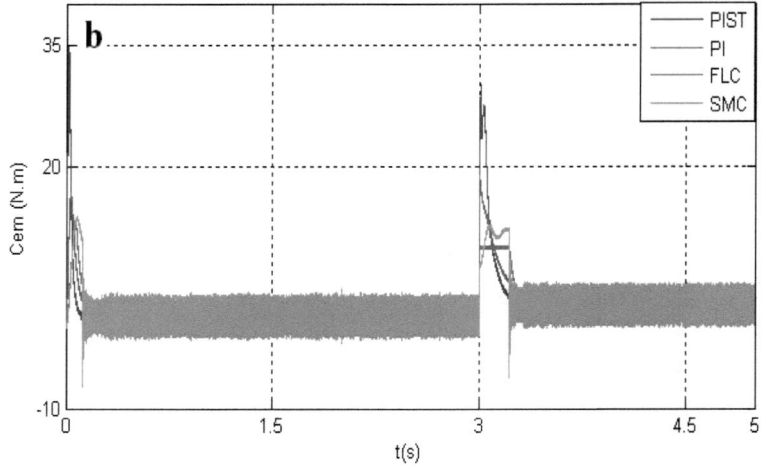

Figure. 18. High tracking responses of the speed (a) and of the electromagnetic torque (b) to change in inertia (case of high speeds, 1st method).

Table 7

Quantitative performances of speed in case of disturbance of the inertia (case of low speeds, 1st method)

| | Controllers | | | |
|---|---|---|---|---|
| | PI | FLC | PIST | SMC |
| Response times (s) to attain 20 (rad/s) | 0.207 | 0.186 | 0.114 | 0.159 |
| Over shoot (%) In case of 20 (rad/s) | 0 | 0 | 1 | 0 |
| Response times (s) to attain 50 (rad/s) | 0.12 | 0.153 | 0.075 | 0.171 |
| Over shoot (%) In case of 50 (rad/s) | 0 | 0 | 1 | 0 |

Table 8

Quantitative performances of speed tracking in case of disturbance of

the inertia (case of high speeds, 1$^{st}$ method)

| | Controllers | | | |
| --- | --- | --- | --- | --- |
| | PI | FLC | PIST | SMC |
| Response times (s) to attain 50 (rad/s) | 0.174 | 0.252 | 0.084 | 0.186 |
| Over shoot (%) In case of 50 (rad/s) | 0 | 0 | 1.82 | 0 |
| Response times (s) to attain 100 (rad/s) | 0.114 | 0.198 | 0.072 | 0.147 |
| Over shoot (%) In case of 100 (rad/s) | 2.3 | 0 | 1.8 | 0 |

## V.1.2 Second Method Simulations

This test aims to evaluate the capacity of the system to follow the reference model as detailed in paragraph IV. As shown in Figure 19, Figure 20, Figure 21, Figure 22, Figure 23 and Figure 24, the response corresponding to the sliding mode controller and fuzzy logic controller is similar to the response of the reference model but the sliding mode controller presents the poor performances to low speeds illustrated by the appearance of the chattering phenomenon. So the system became insensible to the variations of the parameters with this method. Also, it is observed that the performance of the self-tuning fuzzy logic controller is the best compared to that of the conventional PI controller and characterized by the small response time compared to the other controllers. The speed quantitative performances are summarized in tables 9, 10, 11, 12, 13 and 14.

Case of the low speeds:

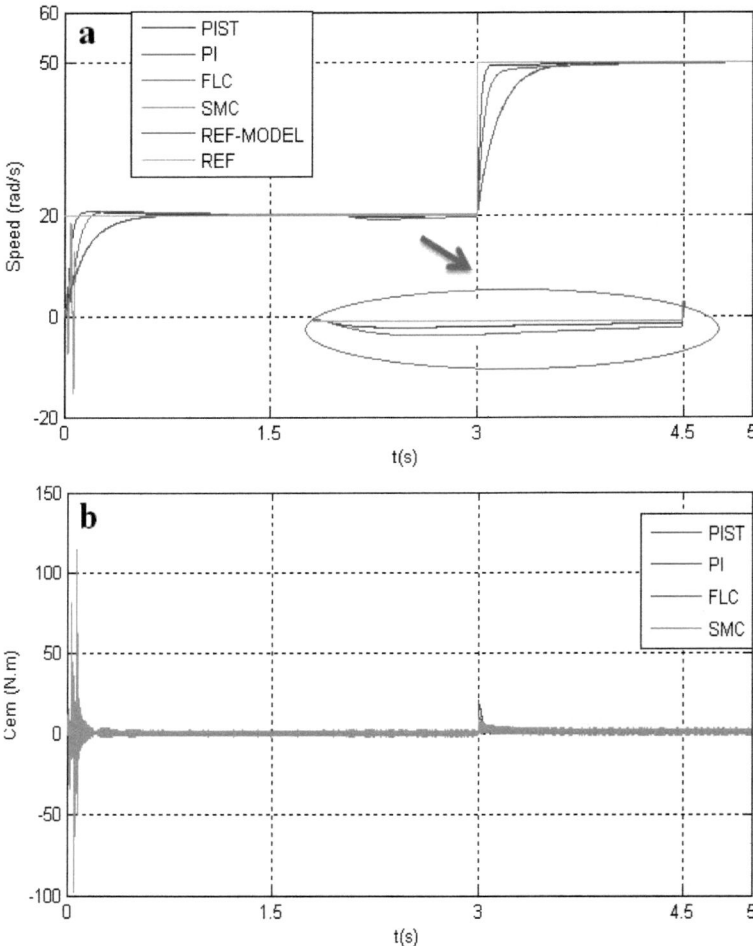

Figure. 19. High tracking responses of the speed (a) and of the electromagnetic torque (b) to change in rotor resistance (case of low speeds, 2$^{nd}$ method).

Case of the high speeds:

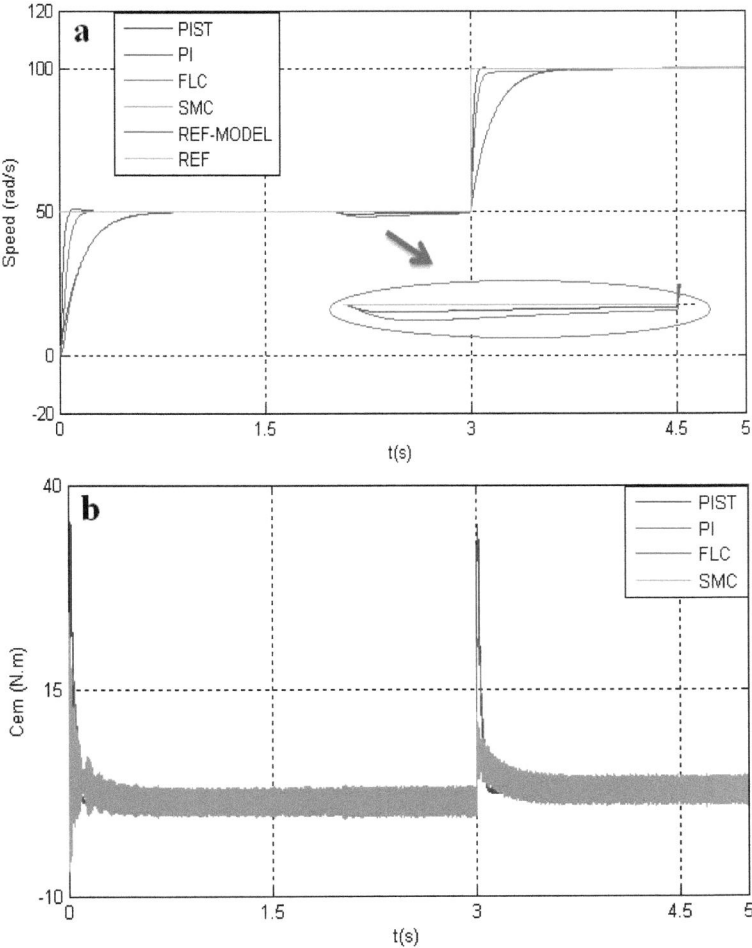

Figure. 20. High tracking responses of the speed (a) and of the electromagnetic torque (b)
to change in rotor resistance (case of high speeds, 2nd method).

Table 9

Quantitative performances of speed tracking in case of disturbance of

the rotor resistance (case of low speeds, $2^{nd}$ method)

| | Controllers | | | | |
|---|---|---|---|---|---|
| | PI | PIST | FLC | SMC | REF-MODEL |
| Response times (s) to attain 20 (rad/s) | 0.297 | 0.144 | 0.453 | 0.459 | 0.459 |
| Over shoot (%) In case of 20 (rad/s) | 0 | 2.15 | 0 | 0 | 0 |
| Response times (s) to attain 50 (rad/s) | 0.042 | 0.087 | 0.225 | 0.225 | 0.225 |
| Over shoot (%) In case of 50 (rad/s) | 0 | 0 | 0 | 0 | 0 |

Table 10

Quantitative performances of speed tracking in case of disturbance of

the rotor resistance (case of high speeds, $2^{nd}$ method)

| | Controllers | | | | |
|---|---|---|---|---|---|
| | PI | PIST | FLC | SMC | REF-MODEL |
| Response times (s) to attain 50 (rad/s) | 0.174 | 0.084 | 0.465 | 0.465 | 0.459 |
| Over shoot (%) In case of 50 (rad/s) | 0 | 1.96 | 0 | 0 | 0 |
| Response times (s) to attain 100 (rad/s) | 0.045 | 0.03 | 0.141 | 0.138 | 0.138 |
| Over shoot (%) In case of 100 (rad/s) | 0 | 0 | 0 | 0 | 0 |

Case of the low speeds:

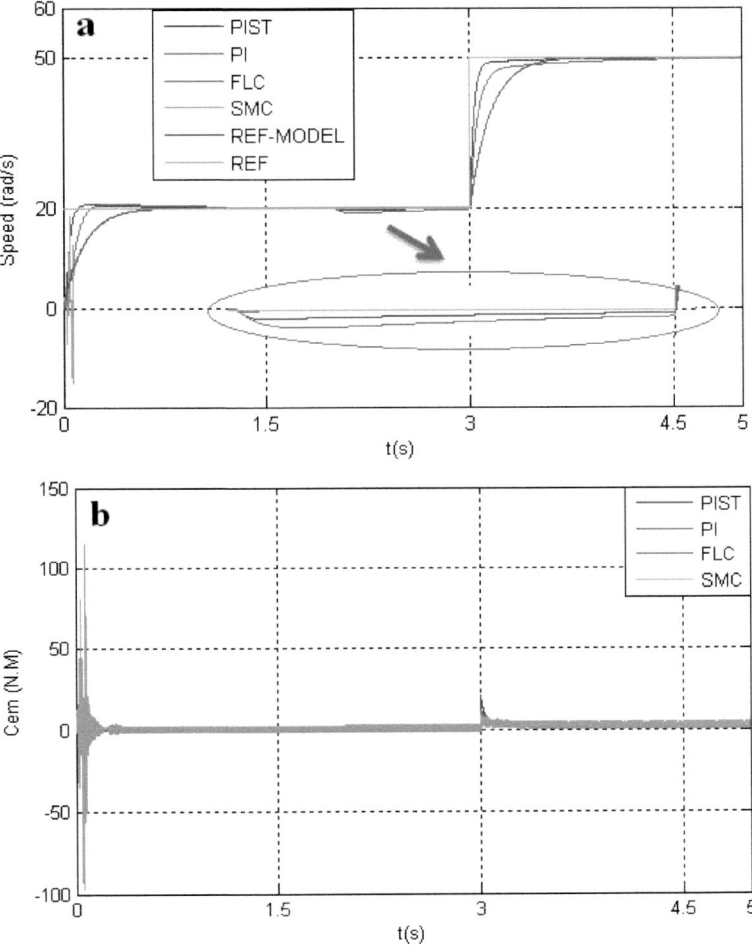

Figure. 21. High tracking responses of the speed (a) and of the electromagnetic torque (b) to change in load torque (case of low speeds, $2^{nd}$ method).

Case of the high speeds:

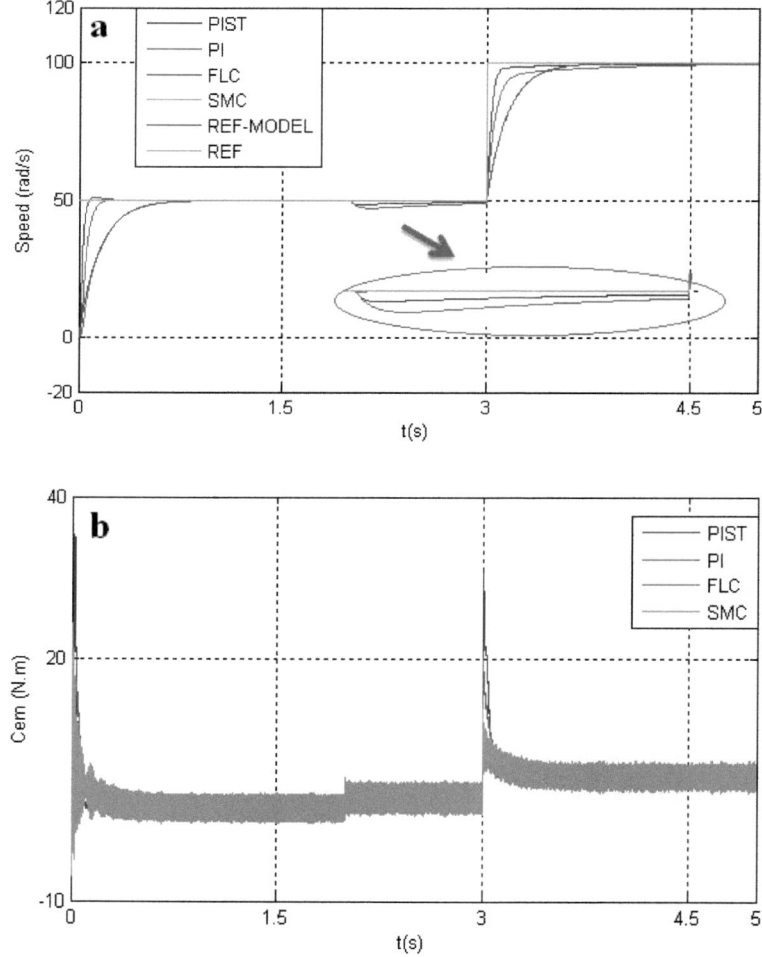

Figure. 22. High tracking responses of the speed (a) and of the electromagnetic torque (b) to change in load torque (case of high speeds, $2^{nd}$ method).

Table 11

Quantitative performances of speed in case of disturbance

of the load torque (case of low speeds, $2^{nd}$ method)

| | Controllers | | | | |
|---|---|---|---|---|---|
| | PI | PIST | FLC | SMC | REF-MODEL |
| Response times (s) to attain 20 (rad/s) | 0.297 | 0.144 | 0.453 | 0.459 | 0.459 |
| Over shoot (%) In case of 20 (rad/s) | 0 | 2.15 | 0 | 0 | 0 |
| Response times (s) to attain 50 (rad/s) | 0.042 | 0.084 | 0.213 | 0.213 | 0.213 |
| Over shoot (%) In case of 50 (rad/s) | 0 | 0 | 0 | 0 | 0 |

Table 12

Quantitative performances of speed tracking in case of disturbance

of the load torque (case of high speeds, $2^{nd}$ method).

| | Controllers | | | | |
|---|---|---|---|---|---|
| | PI | PIST | FLC | SMC | REF-MODEL |
| Response times (s) to attain 50 (rad/s) | 0.174 | 0.084 | 0.465 | 0.459 | 0.459 |
| Over shoot (%) In case of 50 (rad/s) | 0 | 1.96 | 0 | 0 | 0 |
| Response times (s) to attain 100 (rad/s) | 0.057 | 0.036 | 0.141 | 0.138 | 0.138 |
| Over shoot (%) In case of 100 (rad/s) | 0 | 0 | 0 | 0 | 0 |

Case of the low speeds:

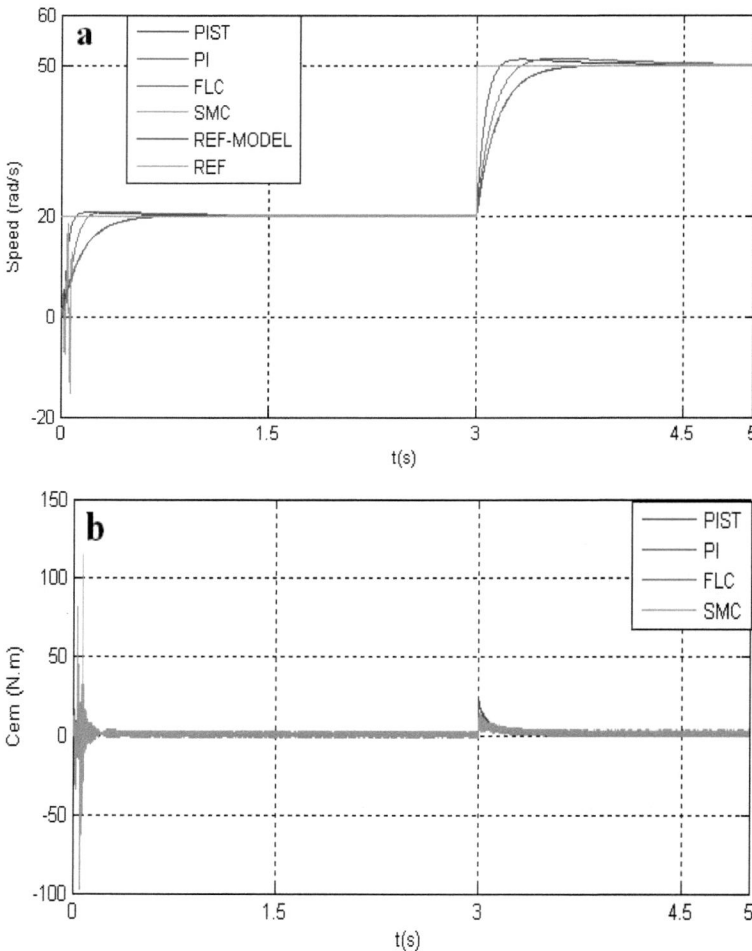

Figure. 23. High tracking responses of the speed (a) and the electromagnetic torque (b) to change in inertia (case of low speeds; $2^{nd}$ method).

Case of the high speeds:

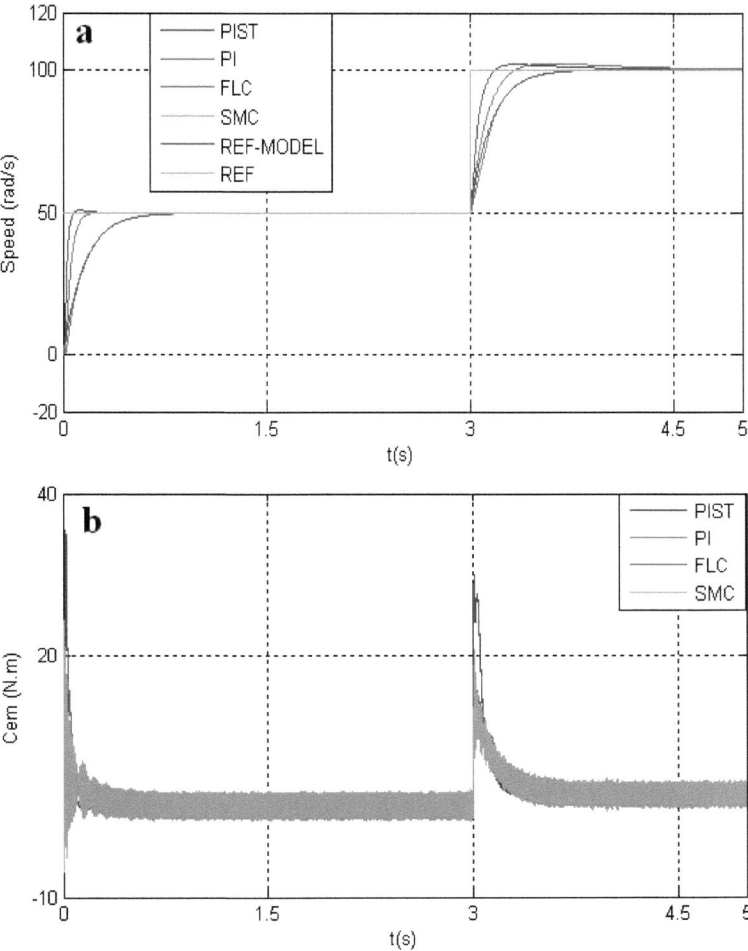

Figure. 24. High tracking responses of the speed (a) and the electromagnetic torque (b) to change in inertia (case of high speeds, $2^{nd}$ method).

Table 13

Quantitative performances of speed in case of disturbance
of the inertia (case of low speeds, $2^{nd}$ method)

| | Controllers | | | | |
|---|---|---|---|---|---|
| | PI | PIST | FLC | SMC | REF-MODEL |
| Response times (s) to attain 20 (rad/s) | 0.297 | 0.144 | 0.453 | 0.459 | 0.459 |
| Over shoot (%) In case of 20 (rad/s) | 0 | 2.15 | 0 | 0 | 0 |
| Response times (s) to attain 50 (rad/s) | 0.159 | 0.084 | 0.225 | 0.225 | 0.225 |
| Over shoot (%) In case of 50 (rad/s) | 0.62 | 1.68 | 0 | 0 | 0 |

Table 14

Quantitative performances of speed tracking in case of disturbance
of the inertia (case of high speeds, $2^{nd}$ method)

| | Controllers | | | | |
|---|---|---|---|---|---|
| | PI | PIST | FLC | SMC | REF-MODEL |
| Response times (s) to attain 50 (rad/s) | 0.174 | 0.084 | 0.465 | 0.459 | 0.459 |
| Over shoot (%) In case of 50 (rad/s) | 0 | 1.96 | 0 | 0 | 0 |
| Response times (s) to attain 100 (rad/s) | 0.108 | 0.066 | 0.195 | 0.141 | 0.141 |
| Over shoot (%) In case of 100 (rad/s) | 2.2 | 1.9 | 0 | 0 | 0 |

## V.2 Simulation results of the second study (Direct torque control)

In order to examine the capacity of the self tuning fuzzy logic controller to compensate the effects of the stator resistance variation, some perturbations on stator resistance have been generated. Two tests have been performed, the first test relates to the low speeds and second test involves high speeds.

The first test aims to evaluate the performances of the self tuning fuzzy logic controller (PIST) at high speeds when the disturbance affects the stator resistance. So, a variation of 100% in the stator resistance is applied at t=2 s and the command is submitted to the reference speed (REF) which varies between -100 rad/s, 0 rad/s and 100 rad/s.

The results simulations show the effects of the stator resistance variation which affects the stability of the machine. As shown in figures 25, 26, 27 and 28, the self tuning fuzzy logic controller rejects the perturbations and maintains the stability of the system. In fact, with a classical PI controller, the errors occur in flux and torque if the variation of the stator resistance is generated. However, the self tuning fuzzy logic controller is characterized by its capacity to adapt the gains parameters, as shown in figures 30 and 31, and therefore minimize the error caused by the variation of the stator resistance.

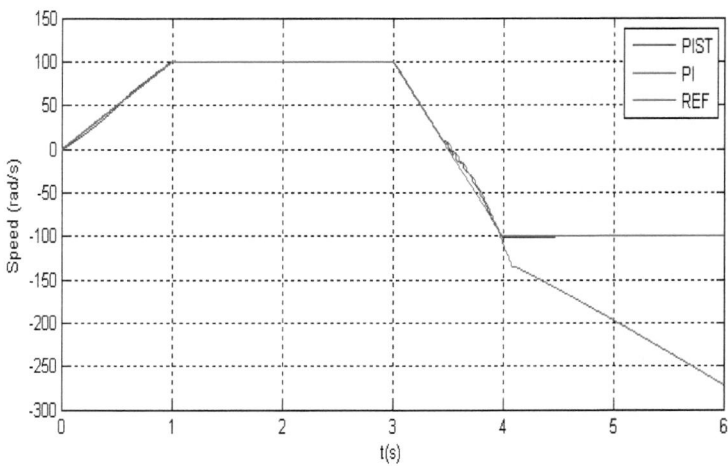

Figure. 25. High tracking response of the speed by using the self tuning fuzzy logic controller (PIST) (case of high speeds)

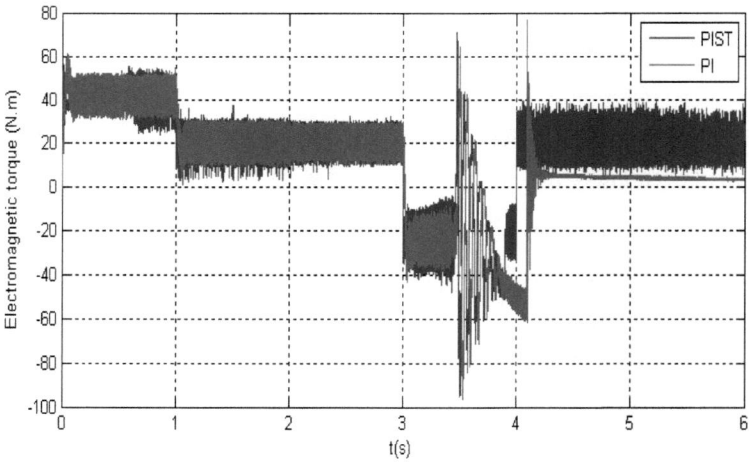

Figure. 26. High tracking response of the electromagnetic torque by using the self tuning fuzzy logic controller (PIST) (case of high speeds)

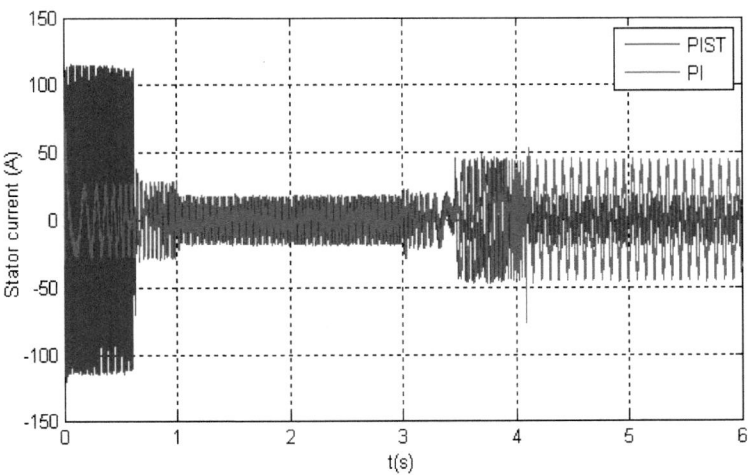

Figure. 27. Evolution of the stator current (case of high speeds)

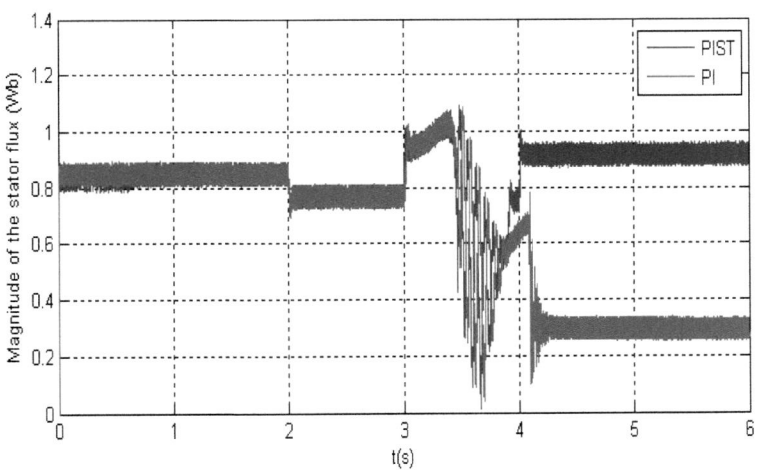

Figure. 28. Evolution of the stator flux magnitude (case of high speeds)

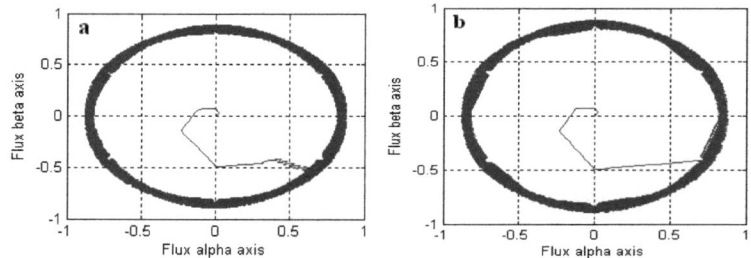

Figure. 29. Evolution of the extremity of the stator flux by using the classical PI controller (a), by using the self tuning fuzzy logic controller PIST (b)

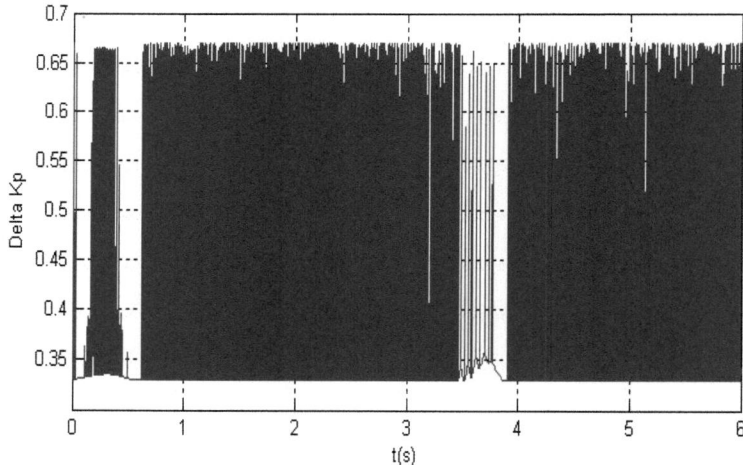

Figure. 30. Evolution of the adaptation factor (ΔKp) (case of high speeds)

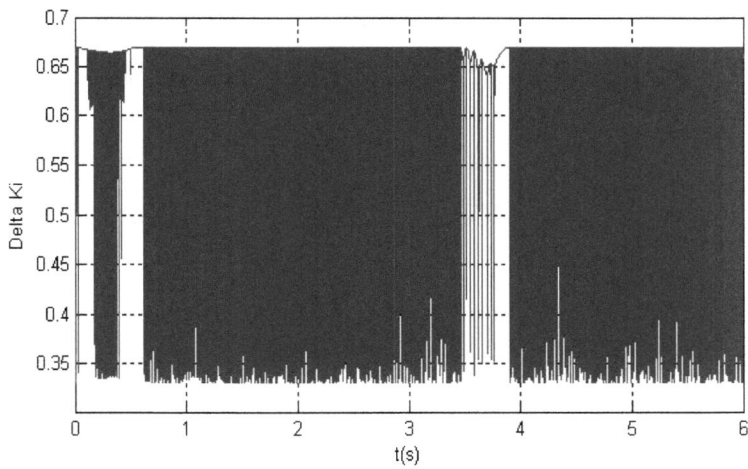

Figure. 31. Evolution of the adaptation factor (ΔKi) (case of high speeds)

The second test has been interested in slow speeds in order to examine the performances of the self tuning fuzzy logic controller in this range of speeds. In the same manner as the previous test, an increase of 100% in stator resistance has been applied at t = 2s. Also, the profile of reference speed (REF) has been varied between 0 rad/s, 10 rad/s and -10 rad/s.

From the simulation results, we note that the self tuning fuzzy logic controller provides very good dynamic responses of the speed, the current, the flux and the torque. In fact, as shown in figures 32, 33, 34 and 35, the cited controller rejects a disturbance perfectly compared to the classical PI controller. However, in case of using the classical PI controller, the stability of the system is maintained with appearance of the oscillations.

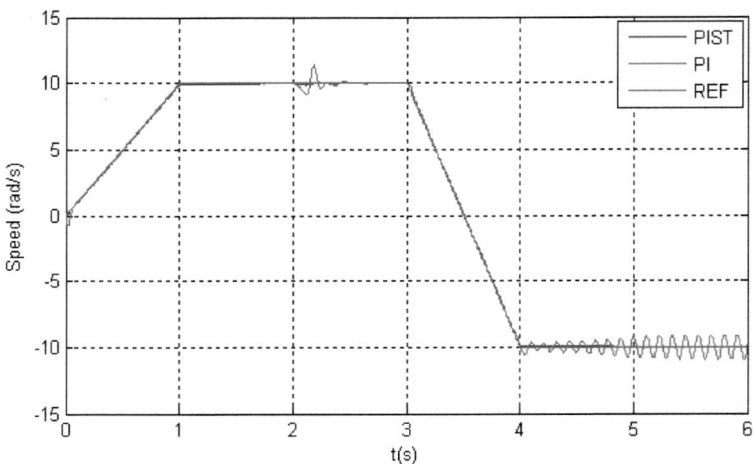

Figure. 32. High tracking response of the speed by using the self tuning fuzzy logic controller (PIST) (case of low speeds)

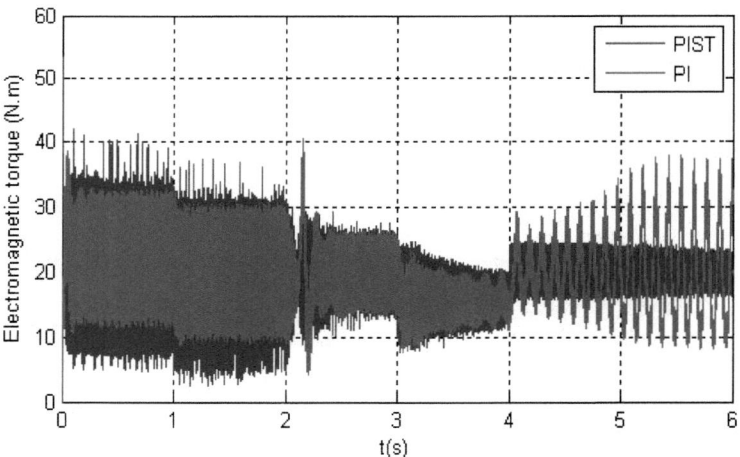

Figure. 33. High tracking response of the electromagnetic torque by using the self tuning fuzzy logic controller (PIST) (case of low speeds)

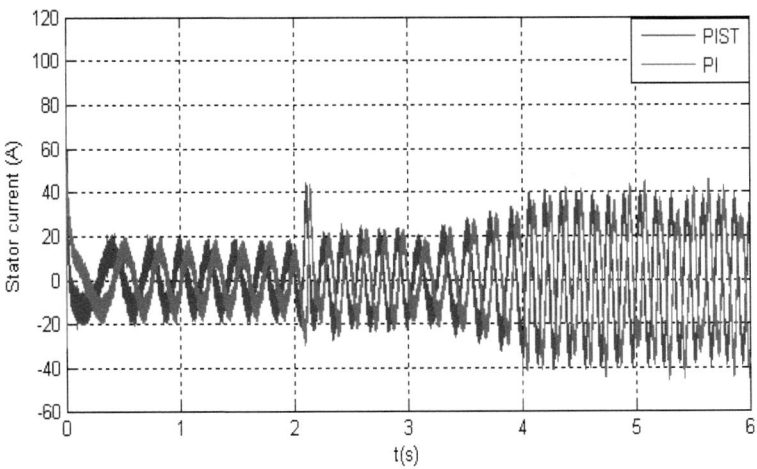

Figure. 34. Evolution of the stator current (case of low speeds)

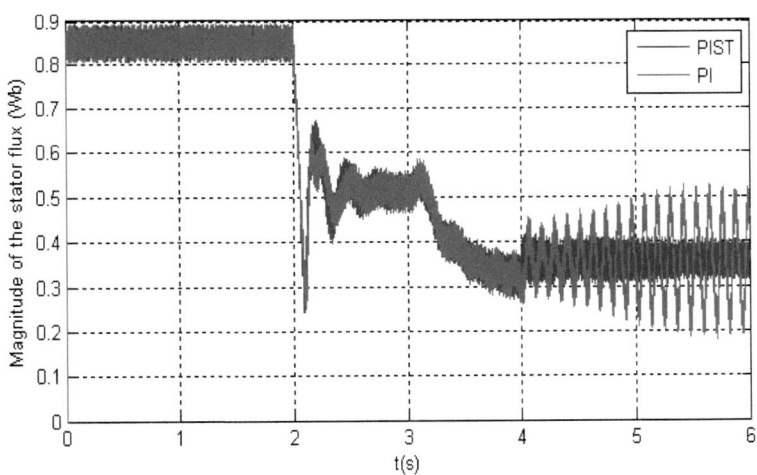

Figure. 35. Evolution of the stator flux magnitude (case of low speeds)

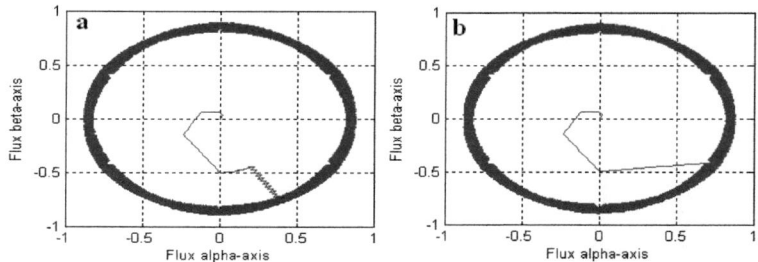

Figure. 36. Evolution of the extremity of the stator flux by using the classical PI controller (a),
by using the self tuning fuzzy logic controller PIST (b) (case of low speeds)

# CONCLUSION

In this work, the author proposes two studies. The first is a comparative study between different control strategies by using different intelligent controllers: sliding mode controller, self tuning fuzzy logic controller and fuzzy logic controller to improve the performances of the indirect rotor field-oriented control by replacing the conventional PI controller with sliding mode controller and self-tuning fuzzy logic controller and the second study presents a solution to improve the performances of direct torque control by elimination of the effect of the stator resistance variation.

According to the simulation results, the sliding mode controller is characterized by a high capacity to reject the disturbance of the machine parameters and the self tuning fuzzy logic controller is characterized by the small response time. Through a series of simulations tests, the sliding mode and the self tuning fuzzy logic controllers present the high performances of speed tracking and disturbance rejection. Also, the self tuning fuzzy controller shows a promising performance to overcome the problem of the sensitivity of the direct torque control to the variation of stator resistance.

The machine used in this work is the induction machine characterized by nominal values: 3 kW, 1400 tr/min, 220/380 V, 12.5/7.2 A, 3 phases, 50 Hz.

The parameters of the used induction machine are summarized in table 1, they were obtained by using the laboratory testing as described in [12].

Table 1
The used IM parameters

| Parameters | Values | Units |
|---|---|---|
| Rotor resistance Rr | 2.68 | Ω |
| Stator inductance Ls | 0.229 | H |
| Rotor inductance Lr | 0.229 | H |
| Mutual inductance M | 0.217 | H |
| Moment of inertia J | 0.046 | Kg.m2 |
| Coefficient of friction f | 0.001 | Kg.m2.s-1 |

# REFERENCES

[1] Z. Zhang, J. Zhu, R. Tang, B. Bai, H. Zhang, Second order sliding mode control of flux and torque for induction motor, *Power and Energy Engineering Conference (APPEEC)*, 2010 Asia-Pacific, March 2010, pp. 1-4, 28-31.

[2] M. Moutchou, A. Abbou, and H. Mahmoudi, Induction machine speed and flux control using vector-sliding mode control with rotor resistance adaptation, *International Review of Automatic Control*, 2012, Vol. 5 Issue 6, pp. 804.

[3] MN. Uddin, TS. Radwan, and A. Rahman, Performances of fuzzy –based indirect vector control for induction motor drive, *IEEA Trans Industry Applications*, Sept/Oct 2002, Vol. 38 Issue5, pp. 1219-1225.

[4] CM. Liaw, and F.J. Lin, Position control with fuzzy adaptation for induction servomotor drive, *IEE Proceedings-Electric Power Applications*, Nov 1995, Vol. 142 Issue 6, pp. 397-404.

[5] K.H. Chao, and C.M. Liaw, Fuzzy robust speed controllers for detuned field-oriented induction motor drive, *IEE Proceedings-Electric Power Applications,* Jan 2000, Vol. 147 Issue1, pp. 27-36.

[6] L. Zhen, and L. Xu, Fuzzy learning enhanced speed control of an indirect field-oriented induction machine drive, *IEEE Trans Control Systems Technology,* March 2000, *Vol. 8 Issue2*, pp. 270-278.

[7] M. Masiala, and A. Knight, Self-tuning speed controller of indirect field-oriented induction machine drives, *Proceedings of the 17th International Conference on Electrical Machines*, Sept 2006, pp. 563568.

[8] M. Chebre, M. Zerikat, and Y. Bendaha, Adaptation des paramètres d'un contrôleur PI par un FLC appliqué à un moteur asynchrone, *Proceedings of the 4th International Conference on Computer Integrated Manufacturing CIP'2007*, November 2007, pp. 03-04.

[9] K. Laroussi, and M. Zelmat, Optimisation floue des parametres du régulateur PI appliqué à un moteur à induction, *Proceedings of the 3[th] International Conference on Sciences of Electronic, Technologies of Information and Telecommunications*, March 2005.

[10] C. Laoufi, A. Abbou, Y. Sayouti, and M. Akherraz, Self tuning fuzzy logic speed controller for performance improvement of an indirect field-oriented control of induction machine, *International Review of Automatic Control, Vol. 6 Issue 4, July 2013*, pp. 464-471.

[11] C. Laoufi, A Abbou and M. AKHERRAZ, Mohammed. Comparative study between several strategies speed controllers in an indirect field-oriented control of an induction machine, *Renewable and Sustainable Energy Conference (IRSEC) 2014 International. IEEE*, 2014, p. 866-872.

[12] A. Abbou, Contribution à l'étude et à la réalisation des stratégies de commande d'un moteur à induction sans capteur de vitesse-simulations et expérimentation, Ph.D. dissertation, Dept. Elect. Eng., Ecole Mohammadia d'Ingénieur, Rabat, Morocco, 2009.

[13] S. Mir, M. Elbuluk and D. Zinger, "PI and Fuzzy Estimators for Tuning the Stator Resistance in Direct Torque Control of Induction Machines," IEEE Transactions on Power Electronics, Vol.13, March 2013.

[14] A. Abbou, H. Mahmoudi and A. Elbacha, "The effect of stator resistance variation on DTFC of induction motor and its compensation," Proceedings of the 14th IEEE International Conference on Electronics Circuits and Systems, Marrakech, Morocco, December 2007.

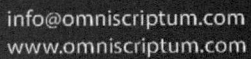

Printed by Books on Demand GmbH, Norderstedt / Germany